Jalapeno Peppers

Production, Processing and Marketing

Roby Jose Ciju

DEDICATION

This book is dedicated to all home gardeners, farmers, commercial growers and all plant-loving souls who have a genuine interest in jalapeno pepper plants and all other chile pepper plants....

CONTENTS

ACKNOWLEDGMENTS

As we know, everything worthwhile is always accomplished through team effort only. I would like to acknowledge the efforts of my staff at agrihortico.com for helping me to republish this version of my book "Jalapeno Peppers" with additional information and relevant images. Hope my readers will find this book very useful in their gardening/farming endeavors...

JALAPENO PEPPERS: AN INTRODUCTION

Jalapeno peppers is an important group of chile peppers. All chile peppers belong to the genus *Capsicum* and the family Solanaceae. Cultivated capsicum includes at least 25 species of chile pepper plants, four of which have been domesticated. They are grown for different purposes such as for vegetables, spices, and condiments. All chile peppers, including jalapeno peppers are tropical and subtropical in their growing habit. Chile peppers are of American origin.

Mainly FOUR species of chile peppers are in cultivation. These are *Capsicum annuum*, *Capsicum pubescens*, *Capsicum chinense* and *Capsicum baccatum* (*C. frutescens*).

Among these, *Capsicum annuum* is the most cultivated chile pepper species. Most of the chile pepper varieties that are available in the market belong to the group *Capsicum annuum*.

1

Jalapeno pepper also belongs to this group.

A detailed account of various chile peppers in cultivation is given below:

Figure 1: A Chile Pepper Plant

Cultivated Chile Peppers in Capsicum Genus

1. *Capsicum annuum var. annuum:* Major cultivated chile peppers in Capsicum annuum group are Bell Peppers, Cayenne Peppers, Cherry Peppers, Pimento Peppers or Banana Peppers, Peperoncino or Italian Sweet Peppers, Jalapeno Peppers, Pasilla Peppers, Poblano Peppers (ancho peppers), Serrano Peppers, Anaheim Peppers (New Mexico Anaheim Peppers) also called 'long green chilies', Hungarian Wax Peppers, Fresno Peppers, Mirasol Peppers, and Cascabel Peppers. Other chile peppers who belong to *Capsicum annuum* group are, Aleppo, Chilaca, Chungyang Red Pepper, Cubanelle, De árbol, Dundicut, Guajillo, Macho, Medusa, Mulato,

Puya, Peter, Santa Fe Grande, Tien Tsin, and Shishito.

Figure 2: Jalapeno Peppers

2. <u>Capsicum baccatum / capsicum frutescens var.</u>
    <u>baccatum:</u> Major cultivated chile peppers in *Capsicum baccatum* group are, Aji (Capsicum baccatum var. pendulum), Peruvian Pepper (Capsicum baccatum var. pendulum), Locoto (Capsicum baccatum var. baccatum), Bird's Eye chili or Thai Peppers, Tabasco peppers and Other cultivars such as African Bird's Eye, Blanco, and Malagueta.

3. <u>Capsicum chinense:</u> Major cultivated chile peppers in Capsicum chinense group are, Rocotillo, Habanero pepper (also called Red Savina Habanero or Scotch bonnet pepper), Naga Jolokia, Trinidad Moruga Scorpion, also known as Trinidad Scorpion Butch T and other cultivars such as Adjuma, Limo, Ají dulce, Datil, Fatalii, Hainan Yellow Lantern Chili, Madame

Jeanette, Mora, and Morita.

4. <u>Capsicum pubescens:</u> One major cultivar belonging to the group *Capsicum pubescens* and is popular among growers is Rocoto Peppers.

In this small book, we will focus on jalapeno peppers only. A detailed account of jalapeno cultivation practices, both conventional growing and organic growing practices for jalapenos, their processing and marketing practices and other relevant information is available in this small book.

ECONOMIC IMPORTANCE OF JALAPENOS

Before explaining production practices, let us have a look at the economic importance of jalapenos.

Jalapeno peppers are mainly used as a vegetable and also as a side ingredient in many exotic food preparations such as pizzas, salads, sandwiches and similar dishes. Jalapenos are excellent food additives and are also used for coloring and flavoring various food preparations. Jalapeno peppers can be grown as ornamental plants as well.

Figure 3: Red and Green Jalapenos

Jalapeno peppers are also a good source of capsaicin, a major antioxidant, presence of which attributes a lot of medicinal properties to jalapenos.

Jalapeno peppers can be processed into a number of valuable processed foods such as nacho sliced peppers, canned jalapenos, smoked jalapenos or chipotles and jalapeno pickles.

So in a nutshell, the economic importance of jalapeno peppers is as follows:

1. As a food additive, flavoring and coloring agent

2. As a vegetable

3. As a spice and condiment

4. As a garnishing and topping agent

5. As an ornamental plant

6. As a rich source of capsaicin, an antioxidant and a health-promoting substance

7. As a medicine, especially a chile tincture

8. As a major processed food item in food processing industries; e.g. chipotles, jalapeno jelly, jalapeno mash, nacho sliced peppers etc

PUNGENCY PRINCIPLE IN JALAPENOS

When it comes to the chile peppers as a whole group, the varieties of chile peppers may be broadly categorized into THREE groups based on its pungency levels. These are: Non-Pungent Chile Peppers, Mildly Pungent Chile Peppers and Highly Pungent Chile Peppers.

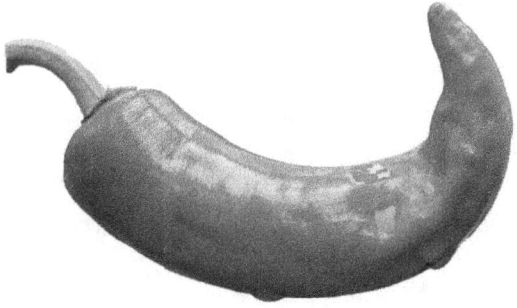

Figure 4: Red Hot Jalapenos

Most of the long narrow type chile peppers such as green chili peppers are highly pungent while short, thick-fleshed chile peppers such as jalapeno peppers are mildly pungent. Bell peppers are non-pungent chile peppers. The pungency of chile peppers is due to the presence of capsaicin, an antioxidant which varies depending on the cultivars and types of the chile peppers.

Jalapenos are mild-pungent chile peppers and are famous for its unique pungency and flavour characteristics. The heat or pungency of jalapeno peppers is measured in Scoville heat units. Scoville heat index of green, fresh jalapeno peppers are about *5,000 SHU* (Scoville heat units).

Scoville Heat Unit (SHU)

Scoville heat unit (SHU) is an organoleptic test based on a subjective dilution-taste procedure developed by Wilbur Scoville, a well-known chemist to measure the pungency or heat factor of chile peppers. SHU is measured in multiples of 100 units where sweet peppers, the mildest of chile peppers is given zero Scoville heat units and the hottest of chile peppers such as Naga Jolokia (Ghost Pepper) and Trinidad Scorpion are given over 1,000,000 Scoville heat units. Heat level of all other chile peppers falls in between these two extremes.

JALAPENO PEPPER PLANTS

Jalapeno peppers are members of chile pepper group.
Scientific name of Jalapeno pepper is *Capsicum annum var.*
annum. It is an herbaceous plant that is grown for their chili-like
fruits. Jalapeno pepper belongs to the family Solanaceae, the
potato family. Jalapeno peppers can be grown as perennial
plants in tropical and subtropical regions. However, for
commercial production, they are grown as annuals to obtain
high yields and to prevent production risks associated with
various disease-pest incidences and climatic variations.

Jalapeno Peppers: Plant Description

Origin: Jalapeno peppers are believed to be originated in the
tropical and subtropical America and later got domesticated in
a region between Mexico and Guatemala. Jalapeno peppers are
commercially produced and consumed throughout the world
with major production centers being in Mexico, China, Turkey,
Nigeria, and the Pacific islands.

Stem: Jalapeno plant is an erect growing herb which reaches to a height of 60–75 cm at maturity. It is a highly branched plant. The plant is primarily self-pollinating. However, cross-pollination may occur sometimes which, by all means, must be avoided by keeping an isolation distance of 150 meters between two jalapeno-growing areas.

Figure 5: A Jalapeno Plant

Roots: Jalapeno plant has a well-developed root system which is extensive with strong and deep taproots.

Leaves: Leaves of the jalapeno plants are elongated, heart-shaped with an attractive, shining green colour. Leaves are arranged alternately on the branches. Flowers are borne solitary in the axils of branches and are produced throughout the year. They are open for 24–30 hours continuously. Flowering begins 1–2 months after transplanting the seedlings in the main fields.

Figure 6: Leaves of Jalapeno Pepper Plant

Adverse environmental conditions during flowering period may decrease fruit set. Therefore, proper care and frost protection are needed during the entire flowering period of jalapeno pepper plants.

Fruits: Fruits are produced on lateral branches and will become ready for harvest about one month after pollination. In other words, fruits can be harvested almost 90 days after transplanting the seedlings. Well-established, well-nourished Jalapeno plants produce fruits all year-round and therefore continuous harvesting of fruits is possible.

Roby Jose Ciju

Figure 7: Fruits of Jalapeno Plants

NUTRITION IN JALAPENO PEPPERS

According to the USDA nutrient database, jalapenos are a powerhouse of nutrients and one of the best low-calorie foods. Jalapeno Peppers are low in calories (29 Kcal/100g) and are rich in dietary fiber (2.8 g/100g) with less fat and zero cholesterol content.

Figure 8: Fresh Green Jalapeno Peppers

Jalapenos are loaded with health-enriching vitamins and minerals also. They are rich sources of Vitamin C (ascorbic acid), which is a powerful antioxidant. They are also rich in vitamin A and therefore are helpful to prevent Vitamin-A deficiencies if consumed regularly. They are a good source of vitamin E, capsanthin, zeaxanthin, and carotenoids and contain very low sodium. Regular consumption of jalapenos contributes to the intake of vitamin C and other nutrients. Jalapenos also contain health-enriching unsaturated fatty acids (MUFA and PUFA). MUFA, mono unsaturated fatty acids are known to have anti-ageing benefits. PUFA, poly unsaturated fatty acids are also considered as healthy fatty acids.

An analysis of 100 gram of edible portion of fresh, raw jalapenos shows following nutrient composition:

Nutrient	Unit	Value per 100.0g
Water	g	91.69
Energy	Kcal	29
Protein	g	0.91
Total lipid (fat)	g	0.37
Carbohydrate	g	6.5
Fiber, total dietary	g	2.8
Sugars, total	g	4.12
Calcium, Ca	mg	12
Iron, Fe	mg	0.25
Magnesium, Mg	mg	15
Phosphorus, P	mg	26
Potassium, K	mg	248
Sodium, Na	mg	3
Zinc, Zn	mg	0.14
Vitamin C	mg	118.6
Thiamin	mg	0.04
Riboflavin	mg	0.07
Niacin	mg	1.28
Vitamin B-6	mg	0.419
Folate, DFE	µg	27
Vitamin A	IU	1078
Vitamin E	mg	3.58
Vitamin K	µg	18.5
Saturated Fatty acids	g	0.092
MUFA	g	0.029
PUFA	g	0.112

Similarly, the nutrient composition of one cup sliced jalapenos and a small jalapeño pepper weighing 14 grams is given below for a comparative analysis:

Nutrient	Unit	cup, sliced 90g	pepper 14g
Water	g	82.52	12.84
Energy	Kcal	26	4
Protein	g	0.82	0.13
Total lipid (fat)	g	0.33	0.05
Carbohydrate	g	5.85	0.91
Dietary Fiber	g	2.5	0.4
Sugars, total	g	3.71	0.58
Calcium, Ca	mg	11	2
Iron, Fe	mg	0.22	0.04
Magnesium, Mg	mg	14	2
Phosphorus, P	mg	23	4
Potassium, K	mg	223	35
Sodium, Na	mg	3	0
Zinc, Zn	mg	0.13	0.02
Vitamin C	mg	106.7	16.6
Thiamin	mg	0.036	0.006
Riboflavin	mg	0.063	0.01
Niacin	mg	1.152	0.179
Vitamin B-6	mg	0.377	0.059
Folate, DFE	µg	24	4
Vitamin A	IU	970	151
Vitamin E	mg	3.22	0.5
Vitamin K	µg	16.6	2.6

HEALTH BENEFITS OF JALAPENOS

As mentioned above, jalapeno peppers are low-calorie foods. A balanced diet is all about right amount of calories. And there are no other foods better than jalapenos to include in your balanced diet. It is recommended as a weight loss diet also. It is believed that jalapeno peppers detoxify carcinogens from the human body system.

They are believed to be used effectively for alleviating medical conditions such as cancers (prostate cancer, skin cancer, lung cancer, stomach cancer), painful joints, toothaches, bronchitis, headaches, ulcers (stomach ulcers, peptic ulcers), arthritis, frostbite and psoriasis. Jalapeño peppers stimulate blood circulation, stop bleeding, and reduce cholesterol. Jalapenos have a soothing effect on the digestive system. Jalapenos are known for its heart stimulant properties also.

A detailed account of various health benefits of jalapeno peppers is given below:

Jalapenos in a Balanced Diet: A right diet is high in nutritional value and provides sufficient calories of energy for bodily functions. All foods, whether carbohydrates, fats or proteins, contain calories, i.e. provide energy. Calorie is a unit of measure of energy. A successful diet program contains right measures of energy. It is all about balance-balance between the calories consumed and the calories required by a body to perform its functions. Surplus calorie consumption results in overweight and insufficient consumption of calorie results in underweight. Both the extremes need to be avoided while planning a balanced diet. Jalapenos are low-calorie foods and hence highly suitable for weight-reducing and health-promoting dietary programs.

Jalapenos are rich in dietary fiber: Dietary fiber is good for human body. Jalapenos are rich source of dietary fiber. High fiber foods is good for weight loss as they take long time to get digested and also make you feel full for a long period of time. High fiber food aids in digestion, cures constipation, lowers blood cholesterol, cleanses the gastrointestinal tract and may reduce the risk of developing diabetes and colorectal cancer.

Jalapenos are rich in Vitamin B6: Vitamin B6 is also known as Pyridoxine. It is essential for fat, carbohydrate and protein metabolism. It also helps in the production of red blood cells (RBCs) and neurotransmitters. Vitamin B6 facilitates proper functioning of estrogen and testosterone hormones in the body. Deficiency symptoms include depression, improper functioning of immune system and sores in mouth. According to US FDA (Food and Drug Administration), daily requirement of vitamin B6 for an average adult is 2mg. A 100 g edible portion of fresh jalapenos supply about 0.419mg of vitamin B6, which is about one-fourth of the daily requirement of this nutrient.

Jalapenos are rich in Potassium (K) mineral: Potassium is an essential mineral that plays an important role in lowering blood pressure. Other major functions of potassium are, carbohydrate metabolism, fluid balance, growth and development, heart function, muscle contraction, nervous system function and protein formation. 100g edible portion of jalapenos contain 248mg of potassium. According to US FDA (Food and Drug Administration), daily requirement of potassium for an average adult is 3500mg. It is clear from the above-mentioned figures that regular consumption of jalapenos contributes to a major portion of our daily requirement of mineral potassium.

Jalapenos are an excellent source of Vitamin C: Vitamin C is also known as ascorbic acid. It is a powerful antioxidant vitamin. Vitamin C helps in absorption of iron and calcium. It increases body's natural immunity. Vitamin C deficiency results in a disease called scurvy. Major symptoms of scurvy are bleeding gum, joint pain, and hair loss. 100g edible portion of jalapenos contain 118.6mg of Vitamin C. It is much higher than the average daily requirement of Vitamin C. According to US FDA (Food and Drug Administration), daily requirement of Vitamin C for an average adult is 60mg.

Jalapenos are moderate source of Vitamin B3 or *Niacin*: Vitamin B3 is also called Nicotinic acid. It is essential for skin health, proper functioning of nerves, and digestion. It also reduces blood cholesterol level and therefore lowers the risk of heart attack. Its deficiency disorder is called Pellagra. Deficiency symptoms include rashes on the skin, dementia and diarrhea. The more severe case of the deficiency leads to death. According to US FDA (Food and Drug Administration), daily requirement of niacin for an average adult is 20mg. A 100 g edible portion of fresh jalapenos supply about 1.28mg of niacin.

Jalapeno peppers are an excellent source of Vitamin K (phylloquinone): Vitamin K is essential for blood clotting and for the formation of strong bones. It is also essential for preventing heart diseases, cancer, and osteoporosis. Vitamin K deficiency results in bleeding gums and bleeding nose. According to US FDA (Food and Drug Administration), daily requirement of Vitamin K for an average adult is 80 micrograms (mcg). A 100 g edible portion of fresh jalapenos supply about 18.5 micrograms of vitamin K.

Jalapeno peppers are moderate source of folate or folic acid or Vitamin B9: This vitamin is essential for energy production from food. It helps in synthesis of nucleic acids and proper functioning of immune system and blood production by facilitating functioning of iron and increasing production of red blood cells (RBCs). It also helps in controlling amino acid metabolism.

Major deficiency symptoms include birth defects in new born babies, diarrhea, hearing loss due to ageing, improper functioning of immune system, weakness, fatigue and headaches. Regular consumption of folic acid helps in slowing down progression of hearing loss with ageing; to prevent birth related defects in new born babies; for protection from cancer, heart diseases, depression and degeneration of body due to ageing; and to prevent memory loss and osteoporosis.

According to US FDA (Food and Drug Administration), daily requirement of folate for an average adult is 400mcg (micrograms). A 100 g edible portion of fresh jalapenos supply about 27 micrograms of folate.

Jalapeno peppers are a good source of Vitamin A: Vitamin A is also known as Retinol. It is essential for eye health and vision. It also strengthens body's natural immune system and is essential for growth and development. Vitamin A is also essential for reproduction, tissue building, and skin and bone health and for the formation of red blood cells. Vitamin A deficiency results in night blindness, and drying of skin and eyes. According to US FDA (Food and Drug Administration), daily requirement of Vitamin A for an average adult is 5000 international units (IU). A 100 g edible portion of fresh jalapenos supply about 1078 IU of vitamin A.

Jalapeno peppers are an excellent source of Vitamin E (alpha-tocopherol): Vitamin E is essential for strengthening body's natural immune system and cardiovascular system. It is a powerful antioxidant vitamin and hence protects the body from heart diseases and cancer. Vitamin E is also required for the formation of red blood cells. Vitamin E deficiency results in weakening of muscular system and nervous system. Other deficiency symptoms include lack of coordination and balance. According to US FDA (Food and Drug Administration), daily requirement of Vitamin E for an average adult is 30 international units (IU). A 100 g edible portion of fresh jalapenos supply about 3.58mg of vitamin E.

FOOD USES OF JALAPENOS

Jalapeno peppers are known as a rich source of vitamins and antioxidant phytochemicals. These nutrients contribute to our health in many ways. There are many medicinal uses of jalapenos. Jalapenos are used in traditional Latin American and African medicinal preparations. Major medical properties of jalapeno are based on its ability to relieve pain from arthritis and migraines. Jalapeno peppers are also good for cough and cold. It also has anti-clotting properties. However, jalapenos are mainly used for a wide variety of food uses.

Figure 9: Food Uses of Jalapenos

Jalapeno peppers are in high demand in bakery industry particularly, in pizza-making food ventures where toppings of pizzas are always filled with jalapeno slices. Nowadays it is hard to imagine pizzas without jalapeno toppings. Likewise, jalapenos are a favorite sandwich filling all across the world.

Jalapenos can be used in foods in several ways. They can be used as a fresh spice condiment, can be stuffed, sliced and cooked, smoked and dried, fried and baked, pickled, and can be made into jellies.

Fresh jalapeno peppers are mainly used as a spice and condiment for flavoring food preparations and also to add 'hotness' to continental food preparations. Pickled and canned jalapenos are also very popular on dining tables. These processed peppers may be added to stews, sauces, salads and other food preparations to add pungency and flavor. Oven dehydrated or smoked peppers (*chipotle*) are a delicacy in Mexico and other Latin American countries.

Some important food preparation methods and recipes with jalapenos are given below:

Food Preparation 1: Stuffed Jalapenos

Freshly harvested large or medium size jalapeno peppers are taken for making stuffed jalapenos. After cleaning the jalapenos in running water, they are dried with a clean towel before removing the seeds. Then the inside portion is filled with the desired filling or stuff. The filling can be prepared with meat or poultry or sea food or cottage cheese or any similar foods.

Food Preparation 2: Pickled Jalapenos

Small jalapenos peppers can be pickled as whole. Large and medium jalapenos can be sliced and pickled in vinegar solution.

Food Preparation 3: Smoked Jalapenos

Fully ripe jalapeno peppers are used for making smoked jalapenos. It is also called chipotles. A detailed account of processing of jalapenos to chipotles is explained elsewhere in this small book.

Food Preparation 4: Meat-Wrapped Jalapenos

Large jalapeno peppers are cleaned and cut longitudinally into two equal slices before filling them with cheese. Meat pieces are also cut into thin longitudinal slices. Jalapeno slices are then wrapped with meat slices before baking them at a high temperature in a hot oven. This dish is a perfect party appetizer.

Food Preparation 5: Jalapeno Jelly

Jalapeno peppers can be used for making jelly also. Fully ripe jalapenos are used for jelly-making. Jelly-making process involves boiling of sliced jalapeno peppers in a sugar-acid solution. Firstly, a clear, strained solution of pectin-containing jalapeno pepper extract is prepared. The pepper extract should be free from all impurities. Then a sugar and acid solution is added into this extract. After that, this mixture is boiled until a transparent and clear jelly is obtained. The jelly should have the original flavor of the jalapenos. Since pectin is the most important constituent of jelly, commercially available pectinic acid, which is water- soluble, may be used. Pectinic acid under suitable conditions forms a gel with sugar and acid. Setting point of pectin is dependent upon the pH of acid and sugar concentration. A Jelmeter is used to determine the pectin content in the jelly. Excessive boiling of jelly should be avoided as it results in a greater inversion of sugar and therefore crystallization of jelly. Excessive boiling also results in the destruction of pectin

Other Uses of Jalapeno Pepper Plants

Jalapeno pepper plants have attractive green foliage and can be grown as garden plants for ornamental purposes as well. Jalapeno plants can be grown for 'Jalapeno extracts' which are used in cosmetics industry as an ingredient in some cosmetic products.

GROWING JALAPENO PEPPER PLANTS

Jalapeno peppers are warm-season crops with a long growing season. Good amount of sunlight and water are essential requirements. They are sensitive to climatic conditions during blossom development and fruiting. A temperature above 32°C may lead to flower abortion. They thrive well under 2–5 cm of rainfall or irrigation per week, depending on the soil type and growth stage of the crop. As a rule, jalapeno peppers require more water after fruiting than before fruiting.

Jalapenos can be grown in any soil. But for maximum yield and quality crop, rich fertile, well-drained, sandy- loam, alkaline soils are ideal. Ideal pH is between 5.0 and 6.5. Jalapeno plants are moderately salt tolerant.

Both open-pollinated and hybrid varieties are available in the market. Difference between these two varieties is given below:

Open-pollinated Hybrids- Seed propagation results in true-to-type generation. Yield is low as compared to hybrids

Hybrids- Seed from hybrid plants are not true-to-type. Growers have to buy fresh seeds every year but yield is high

Commercial varieties of jalapeno peppers are available from seed companies from Asia and North America. A list of major commercial varieties of jalapeno peppers is given below:

1. Jalapa- A popular variety in Asia

2. Waialua- Resistant to root-knot nematodes and to bacterial wilt; a popular variety among home-gardeners in Hawaii

3. Jalapeno M- A popular variety in Asia

4. Mitla- A popular variety in Asia

5. TAM Mild Jalapeño- A mild variety of jalapeno released by an American university

Where to Find Technical Information?

In USA, seed availability-related information is available from USDA while growers located in the Asian continent may contact the Asian Vegetable Research and Development Center for further information.

Parameters for Varietal Selection

Major parameters that are used for varietal selections are,

1. Target market (e.g. local vs. global; processing vs. fresh)

2. Results of the local variety trials

3. Desired horticultural traits that are required to meet the needs of the target market (such as yield, color, shape, and flavor)

4. Adaptability to a particular location

5. Tolerance or resistance to fungal, bacterial and viral diseases

6. Tolerance to root-knot nematodes, insects and pests

Propagation: Jalapeno pepper plants are mainly propagated through seeds. Favorable temperature for seeds germination is 15–30°C.

Raising of Seedlings: Seedlings are raised in well-prepared nursery beds. Nursery beds are prepared in a well-shaded area and seeds are either broadcasted or sown in rows in nursery beds. Germination process is slow and it may take up to 12 weeks for the seeds to germinate. Regular light watering is necessary to keep the nursery bed moist always. Moist soil facilitates quick germination of the seeds and seedlings will become ready for transplanting in the main field when they reach up to 6-8 weeks old.

Transplanting: Six to eight week old seedlings or seedlings with four or more true leaves are transplanted indwell-prepared fields where sunlight is abundantly available. Before planting the seedlings, field is well prepared by mixing bulk loads of FYM (farm yard manure or barn yard manure) or organic compost in the topsoil to enrich the soil fertility. Since seedlings are prone to direct frost, care should be taken to minimize frost damage.

Planting Time: Seedlings are transplanted after the frost is over. That is, ideal time for transplanting is from February to March for spring crop.

Land Preparation: Primary tillage and land preparation operations include chopping prior crop residues, ploughing and tilling the land, leveling, and preparation of field beds. These operations are generally performed from December through January in tropical and sub-tropical regions. FYM (farm yard manure) or compost may be applied @ 25 tons/hectare (10-12 tons/acre) at the time of land preparation to improve the fertility of the top soil. Other practices that can be adopted for improving soil fertility are use of organic mulches; rotations with green manures or other cover crops and adding biofertilizers that help improve beneficial microbial associations. For example, mycorrhizal associations with the plant roots increase the uptake of phosphorus (P) in P-deficient soils.

Spacing: A spacing of 12 inches apart between the plants within a row and a spacing of 18 inches between two rows is recommended for intensive commercial production.

For other purposes such as vegetable gardening and home gardening purposes, a spacing of at least 24 inches between two plants and 3 feet between two rows may be practiced.

Adequate spacing between two plants and between two rows/ridges is necessary because of the vigorous vegetative growth habit of jalapeno plants. An individual plant may grow up to 3 feet high upon maturity.

Planting Density: Recommended planting density is approximately 14,500 – 15,000 transplants per acre (approx. 30,000/ha).

Flowering: Flowering begins approximately after 8 weeks of transplanting.

Watering or Irrigation: Watering depends on the prevailing climatic conditions and soil types. Watering is done on every alternate day when climate is mild and daily watering is recommended when climate is dry. Light sandy soils will require more frequent irrigations than heavier soils. Major point is to keep the soil moist always and avoid overwatering. Excess moisture in the root zone may lead to rotting of the roots while lack of adequate moisture results in loss of flowers and fruit drop.

Weeding: Frequent manual weeding is recommended for jalapeno plants. Care should be taken to keep the field free of weeds as vigorous-growing weeds may compete for nutrition with the jalapeño plants. Black plastic mulch may also be used for weed control. If weed infestation is very high, chemical weed control by using recommended herbicides may be practiced.

Staking: Staking using either wooden logs or bamboo sticks or similar strong supports is advised for areas that experience heavy winds. Staking is also recommended for growing hybrid varieties that produce heavy fruit loads.

Fertilizer Schedule: Fertilizer schedule for Jalapeno plants is as given below:

At the time of land preparation:

1. 25 tons of FYM or compost /hectare
2. For better results, apply adequate quantities of neem cake or any available oil cakes along with FYM

Soon after transplanting:

A "starter" fertilizer solution is recommended soon after transplanting to promote root growth. As a rule, this starter should be low in nitrogen and proportionately higher in phosphorus. A chemical starter solution may consist of 350 gram of a 10-52-17 fertilizer per 50 liters of water (or about 70 gram per 10 liters of water). Application rate is 250 ml of stock solution per plant. An organic starter solution may consist of about 800 ml of fish-based fertilizer or similar organic fertilizers in 50 liters of water.

Recommended nutrient requirements: NPK @120:80:50kg /ha

1/3 rd of N and full quantities of P and K are applied as basal dose and 2/3 rd of N is applied in two equal doses 30 and 60 days after transplanting

Precautions to be taken: While fertilizing the plants, care must be taken not to apply excess nitrogen as it may cause blossom-end rot. If traditional irrigation methods are followed, after fertilizer application, main field should be irrigated through furrow system of irrigation. If drip irrigation system is followed, fertigation may be practiced; that is, fertilizers may be dissolved in water and supplied through drip irrigational water.

Disease-Pest Management: Major pests of jalapeno plants are cutworms, aphids, whiteflies, flower thrips, mites, and the pepper weevil. Root-knot nematodes are also found affecting jalapeno pepper plants. Major diseases are bacterial wilt, bacterial spot, powdery mildew, damping-off, another root-rots caused by Phytophthora or Pythium. Virus diseases are also found affecting jalapeno plants.

Control of Pests and Diseases: The best control measure is to keep a healthy ecosystem. Key components of a healthy agro-ecosystem include, proper land preparation prior to planting;

improving soil fertility by the use of organic manures and fertilizers, organic mulches, and rotation with cover crops; selection of varieties adapted to the location; proper irrigational management to prevent drought or water logging; use of windbreaks and/or intercrops to minimize damage caused by excessive winds; intercropping with compatible crops and understanding about pest life-cycles.

ORGANIC GROWING OF JALAPENO PLANTS

Growing plants organically is becoming a healthy trend nowadays because of the eco-friendly growing practices adopted in organic agriculture. The products of organic agriculture are extremely safe to consume without fearing any dangers of pesticidal residues.

The key areas of organic agriculture is the use of planting materials of organic origin, the use of organic manures and fertilizers for plant nutrition, adopting integrated pest management technology for controlling insects and pests and integrated disease management for controlling disease and also using a lot of beneficial cultural practices such as crop rotation, companion planting, using trap crops etc for maintaining a healthy ecosystem in the growing fields.

Organic Growing of Jalapenos

Major steps involved in organic growing of jalapenos are:

1. Choosing planting materials

2. Preparing soil and enhancing soil fertility

3. Integrated Pest Management

4. Integrated Disease Management

5. Other Cultural Practices such as Crop Rotation, Companion Planting, Using Trap Crops etc

Choosing Plant Materials: In organic production, we need to use planting materials of organic origin. Jalapenos are propagated by seeds and seedlings of organic origin may be purchased from a certified organic plant nursery.

Figure 10: Raising Seedlings

Preparing Soil: For growing jalapenos, as we all know, soil is the major growing medium for plants. So soil should be fertile with all essential plant nutrients to support a healthy plant growth. Generally, soil fertility is enhanced by the addition of fertilizers.

Before understanding about various fertilizers, we can look at various soil sterilization procedures. In organic farming, seedlings are planted in a sterilized soil medium so that soil-borne plant pathogens can be prevented. Soil sterilization can be done by soil solarization method. Soil solarization is a method of trapping solar energy within the soil by covering the soil with a transparent polyethylene cover for a certain period of time. This practice kills all soil-borne plant pathogens.

After soil sterilization, methods need to be adopted to enhance soil fertility. This is accomplished through the addition of fertilizers.

Now, what is a fertilizer? Fertilizer is the term used to refer any material that provides essential nutrients for plant growth when applied externally, either by mixing with soil or by dissolving in the water. A fertilizer is rich in essential plant nutrients. Fertilizer may be either in solid form or in liquid form.

Fertilizer application is an important cultural practice while growing plants. It is the practice of adding fertilizers to the soil or any other growing media through direct broadcasting; or by direct mixing with the soil or through spraying of water-soluble fertilizers. There are three types of fertilizers and these are Organic Fertilizers, Biofertilizers and Inorganic Fertilizers or Chemical Fertilizers.

In organic farming practices, we do not use chemical fertilizers. We generally used organic manures and fertilizers as well as biofertilizers.

Now let us understand in detail about organic manures and organic fertilizers...

Any organic matter used as organic fertilizer is known as organic manure. Manures contribute to the fertility of the soil by adding organic matter and nutrients. There are three main classes of organic manures used in soil management: animal manures; compost and plant manures.

Animal Manures

Common forms of animal manure include farmyard manure (FYM) or farm slurry (liquid manure). FYM also contains plant material (often straw), which has been used as bedding for animals and has absorbed the feces and urine. Agricultural manure in liquid form, known as slurry, is produced by more intensive livestock rearing systems where concrete or slats are used, instead of straw bedding. Sheep manure is high in nitrogen and potash while animal manures like pig manures are relatively low in nitrogen and potash. Horse manure may contain lots of weed seeds, as horses do not digest seeds the way that cattle do. So using horse manure may result in a lot of weed infestation in the fields. Chicken manure, even when well rotted, is very concentrated and should be used sparingly for growing plants. Animal manures may also include other animal products, such as wool shoddy (and other hair), feathers, blood and bone etc.

Compost

Compost is a decomposed plant matter that is used as an organic fertilizer, soil amendment and as a source rich in humus. Compost is used in organic farming, gardening, landscaping, horticulture, and agriculture. It may be used as a natural pesticide for soil and is useful for soil erosion control, soil reclamation, and waste land management.

Figure 11: Compost

Composting is a natural microbiological process of decomposing organic matter into humus and minerals. Microorganisms that aid composting process include bacteria, actinomycetes, fungi, protozoa and earthworms. Composting process can be accelerated by shredding the leaves and adding extra nitrogen on the shredded leaves. Adding water also hastens composting process. Proper aeration must be ensured by regularly turning the mixture. The composting process is entirely dependent on micro-organisms to break down organic matter into compost. Presence of a healthy microbial community is essential for rapid decomposition process. Lack of a healthy microbial community makes composting process very slow. With the proper mixture of water, oxygen, carbon, and nitrogen, micro-organisms work faster to break down organic matter to produce compost.

Shredding leaves make a homogenous compost mixture

with smaller particles. Smaller particles decompose faster as there are more surfaces for the microbes to work on. Particles should not be very small as very small particles may compact and restrict oxygen availability. A blend of small and large particles will be most efficient.

Compost can be prepared by using traditional process or by using modern composting techniques. Traditional composting is a slow process and it takes about one year or more to finish the composting process. It is simple, sometimes requires simply piling up waste outdoors or in pits. Modern composting uses containers or composting bins for composting process. It is a multi-step, closely monitored process and it uses more homogenized pieces in the compost. It uses measured inputs of water, air and carbon- and nitrogen-rich materials and is a rapid process. Modern composting process takes about 2 to 3 weeks to complete.

Major considerations while preparing compost are: Carbon, Nitrogen and Oxygen. Microbial oxidation of carbon produces the heat required for decomposition process. Nitrogen is necessary as microbes grow and reproduce consuming nitrogen-rich plant materials. Oxygen is also an important element as microbes oxidize carbon using the oxygen present in the compost pile. Water should also be added as presence of water maintains life activities of microbes within the compost pile.

Optimum temperature should be maintained throughout the composting procedure. Higher temperature may kill the microbes. At low temperatures, they may remain inactive. The most efficient composting occurs with a carbon: nitrogen mix of about 30 to 1. Plant and animal materials have both carbon and nitrogen. If nitrogen is less in a compost pile, use urea fertilizer or other nitrogen-rich materials.

Standard uses are:

- 1 lb. urea to 1 cubic yard. leaves
- 6 lb. urea to 1 cubic yard. wood clippings
- 5 parts leaves to 1 part manure
- Dried blood meal, alfalfa meal at the rate of 2 cups to a wheelbarrow load of brown leaves or other carbon rich wastes such as shredded paper

Moisture level should always be maintained at about 50%. Compost pile should always be moist and should never be kept dry. Overwatering should be avoided. Microbes need sufficient oxygen for decomposing compost. Since composting is an aerobic process, absence of oxygen causes anaerobic conditions causing a bad odor and partial decomposition of the compost. Hence it is essential that compost pile must be turned at regular intervals to facilitate aeration. Restrict size of the pile to no more than 5 ft. high and 5 ft. wide. Compost pile of 4x4 ft makes an ideal size. Optimum temperatures are between 100° and 140°F as higher temperatures than that may kill microbes. Composting at the center of the pile is complete when temperatures within the pile drop below 100° F. Once composting at the center is complete, turn the pile, putting outside edges inside and allow it to compost more.

Good compost will be dark in color; friable and porous and it will have an earthy smell. It is an excellent source of humus and plant nutrients and has good water holding capacity. Compost is used as an additive to soil and is used as a tilth improver, supplying humus and nutrients. It provides a rich growing medium and acts as a porous, absorbent material that holds moisture and soluble minerals, providing the support and nutrients in which plants can flourish. Compost may be mixed with soil, sand, grit, bark chips, vermiculite, perlite, or clay granules to produce loam.

Vermicompost

Vermicompost is the product of composting utilizing various species of worms. Red wigglers, white worms, and earthworms are used for vermicomposting. Vermicompost is a heterogeneous mixture of decomposed vegetable or food waste, bedding materials, and Vermicast. Vermicast, also known as worm castings, worm humus or worm manure, is the end-product of the breakdown of organic matter by species of earthworms. Red wigglers are recommended by vermiculture experts as they voraciously feed on compost pile and breed very quickly. Vermicompost contains water-soluble nutrients and is a nutrient-rich organic fertilizer and soil conditioner.

Figure 12: Vermicompost

Green Manures

Green manures are crops grown for the express purpose of ploughing them in, thus increasing fertility through the incorporation of nutrients and organic matter into the soil. Leguminous plants such as clover are often used for this, as they fix nitrogen using Rhizobial bacteria in specialized nodes in the root structure. Leguminous cover crops are also grown to enrich soil as a green manure through nitrogen fixation from the atmosphere as well as phosphorus (through nutrient mobilization) content of soils. Other types of plant matter used as manure include the contents of the rumens of slaughtered ruminants, spent hops (left over from brewing beer) and seaweeds.

Figure 13: Clover, A Green Manure Crop

What Are Organic Fertilizers?

Organic fertilizers are naturally occurring fertilizers and mineral deposits. Organic manures are just one form of organic fertilizers. Other examples of organic fertilizers are vermicompost or worm castings, compost, seaweed, guano, naturally occurring mineral deposits (e.g. saltpeter) etc.

Processed organic fertilizers include compost, blood meal, bone meal, humic acid, amino acids, and seaweed extracts. Other examples are natural enzyme digested proteins, fish meal, and feather meal. Decomposing crop residue (green manure) is also used as an organic fertilizer.

Mined powdered limestone, rock phosphate and sodium nitrate, are inorganic (not of biologic origins) compounds but are approved for usage in organic agriculture in minimal amounts.

Figure 14: Green Manures

Organic fertilizer nutrient content, solubility, and nutrient release rates are typically much lower than mineral (inorganic) fertilizers. All organic fertilizers are classified as 'slow-release' fertilizers, and therefore cannot cause nitrogen burn. Organic fertilizers are low-cost as compared to inorganic fertilizers. An organic fertilizer improves the biodiversity (soil life) and long-term productivity of soil and also increases the abundance of soil organisms by providing organic matter and micronutrients for organisms such as fungal mycorrhiza, which aid plants in absorbing nutrients.

Initially organic fertilizers may not be as effective as inorganic fertilizers but application of organic fertilizers become as effective as chemical fertilizers over longer periods of continuous use.

Some of the major advantages of organic fertilizers are as

follows:

1. Nitrogen supplying organic fertilizers contain insoluble nitrogen and act as a slow-release fertilizer

2. Increase physical and biological nutrient storage mechanisms in soils, mitigating risks of over-fertilization

3. Mobilize existing soil nutrients, so that good growth is achieved with lower nutrient densities while wasting less

4. Release nutrients at a slower, more consistent rate, helping to avoid a boom-and-bust pattern

5. Help to retain soil moisture, reducing the stress due to temporary moisture stress

6. Improve the soil structure

7. Help to prevent topsoil erosion

Some of the major disadvantages of organic fertilizers are as follows:

1. Organic fertilizers may contain pathogens and other disease causing organisms if not properly composted

2. Nutrient contents are very variable and their release to available forms that the plant can use may not occur at the right plant growth stage

3. Organic fertilizers are comparatively voluminous and can be too bulky to deploy the right amount of nutrients that will be beneficial to the plants

4. The nutrients in organic fertilizer are both more dilute and also much less readily available to plants

5. As a dilute source of nutrients when compared to inorganic fertilizers, transporting large amount of fertilizer incurs higher costs, especially with slurry and manure

6. The composition of organic fertilizers tends to be more complex and variable than a standardized inorganic product

7. More labor is needed to compost organic fertilizer, thus increasing labor costs

8. Reduce external inputs of pesticides, energy and fertilizer, at the cost of decreased yield

Plant Nutrition: An Overview

Plants absorb nutrients from the soil or the atmosphere, or from water. Carbon and oxygen are absorbed from the air while other nutrients including water are obtained from the soil. Three ways of nutrient uptake are Simple Diffusion, Facilitated Diffusion and Active Transport. Plants absorb essential elements from the soil through their roots and from the air (mainly consisting of carbon and oxygen) through their leaves. There are 17 essential plant nutrients grouped into two categories: macro-nutrients and micronutrients.

Major macro-nutrients are carbon, hydrogen, oxygen, nitrogen, phosphorus, potassium, calcium, magnesium, sulfur and silicon. Major micro-nutrients are boron, copper, chlorine, iron, manganese, molybdenum and zinc. Trace elements like sodium, nickel and cobalt may also be needed in certain circumstances.

Macronutrients are taken by plants in large quantities and are present in plant tissue in quantities from 0.2% to 4.0% dry weight while micronutrients are needed in small quantities and are present in plant tissue in quantities measured in parts per million, ranging from 5 to 200 ppm, or less than 0.02% dry weight. In the absence of an essential element, the plant is unable to complete a normal life cycle. An essential element is a part of some essential plant constituent or metabolite. An element present at a low level may cause deficiency symptoms, while the same element at a higher level may cause toxicity. Deficiency of one element may present as symptoms of toxicity from another element. An abundance of one nutrient may cause a deficiency of another nutrient. A lowered availability of a given nutrient may affect the uptake of another nutrient. The root, especially the root hair, is the most essential organ for the uptake of nutrients.

Now let us have a look at various functions of plant

nutrients in detail...

- Carbon: It is backbone of many plants biomolecules, including starches and cellulose and is a part of the carbohydrates that store energy in the plant

- Hydrogen: It is necessary for building sugars and building the plant and is obtained from water. It is necessary for electron transport chain in photosynthesis and for respiration

- Oxygen: It is necessary for cellular respiration

- Nitrogen: It determines green color and density in plant and is needed for chlorophyll, which is needed for photosynthesis. It also improves plant's ability to resist disease and tolerate effects of heat, cold, and drought. Major deficiency symptom includes yellowing of leaves called chlorosis

- Phosphorus: It helps plants hold and transfer energy for metabolism and also affects cell division, root development, and flowering. Deficiency symptom includes purple coloring of leaves or stems

- Potassium: It activates enzymes and regulates opening and closing of stomata. It also regulates water uptake by root cells

  - Calcium: It regulates transport of other nutrients into the plant and is involved in the activation of certain plant enzymes

- Magnesium: It is important part of chlorophyll and is important in the production of ATP through its role as an enzyme cofactor
- Sulphur : It is a structural component of some amino acids and vitamins and is essential in the manufacturing of chloroplasts
- Silicon: Silicon is deposited in cell walls and contributes to its mechanical properties including rigidity and elasticity
- Iron: It is necessary for photosynthesis ; Present as an enzyme cofactor in plants
- Molybdenum: It is a cofactor to enzymes important in building amino acids
- Boron: It is important for binding of pectins in the RGII region of the primary cell wall and also it plays a significant role in sugar transport, cell division, and synthesizing certain enzymes
- Copper: It is important for photosynthesis; Involved in many enzyme processes; Necessary for proper photosynthesis; Involved in grain production; Involved in the manufacture of lignin (cell walls)
- Manganese: It is necessary for building the chloroplasts
- Zinc: It is required in a large number of enzymes and plays an essential role in DNA transcription

- Chlorine: It is necessary for osmosis and ionic balance; Plays a role in photosynthesis

- Nickel: It is essential for activation of urease, an enzyme involved with nitrogen metabolism and it can substitute for Zinc and Iron as a cofactor in some enzymes

- Sodium: It is involved in the regeneration of phosphoenolpyruvate in CAM and C4 plants and it can also substitute for potassium in some circumstances

- Cobalt: It is essential for legumes for nitrogen fixation and it can substitute for molybdenum

We all know that under-nourishment of plants may lead to poor growth and development. But over-nourishment is also not good for plants.

Over-fertilization or adding extra fertilizer doses to plants may lead to nutrient toxicity. Five types of nutrient toxicity are Chlorosis (yellowing of plant tissue caused by a shortage of chlorophyll synthesis), necrosis (death of plant tissues), accumulation of Anthocyanins (production of a purple or reddish colorization of foliage and/or stems), lack of new growth (stunting or reduced growth) and stunted new growth.

So the thumb rule in fertilizer application is as follows:

Thumb rule in fertilizer application: Thumb rule is to apply right quantities of fertilizers at the right time using a right method. That is,

- Apply right quantities of fertilizers
- Apply fertilizer when the plants can best use the nutrients
- Apply small amounts of fertilizer frequently
- Be careful not to over fertilize

Integrated Pest Management

Another important organic growing practice is the use of IPM for Pest management. A detailed description of various IPM practices is given below:

IPM: Integrated Pest Management or IPM is a holistic approach for the control of pests using all available pest control practices such as cultural control, mechanical control, biological control and chemical control of pests while ensuring the safety of the environment. There are mainly chemical, biological, cultural and mechanical control of insects and pests in IPM.

Chemical Control: Chemical control of pests is accomplished by using chemical pesticides. Various pesticides are insecticides for controlling insects, rodenticides for controlling rats and other rodents, acaricides for controlling ants and termites, miticides for controlling mites and fungicides for controlling fungi.

Biological Control: Biological control of pests is done by using predators (natural enemies) and biopesticides. In organic growing practices, biological control of pests is used extensively. Biological pest control uses natural enemies of the pests to suppress their growth and multiplication. Natural enemies may include predators, fungi and bacteria, and nematodes.

Figure 15: Ladybird Beetles, Natural Enemies of Aphids

Mechanical Control: Mechanical control of pests uses mechanical devices to trap and destroy insects and pests. Pests can be controlled by using mechanical devices such as by covering the greenhouse doors and ventilators with wire meshing, by using light traps for attracting and destroying pests, and by using sticky cards of blue (for thrips) and yellow (aphids and whiteflies) colors for pest monitoring.

Cultural Control: Cultural control of pests is accomplished through daily monitoring and sampling of pests for pest identification and thereafter adopting appropriate cultural practices for controlling these pests. Pest infestation can be controlled by adopting various cultural practices such as use of sterilized growing media or soil for plant growth, by avoiding overwatering and moist growing media to prevent the infestation of fungal molds, by clearing the growing plots of all diseased and pest infested plants and debris and by providing good ventilation for all plants.

Sometimes insect-pests may develop resistance to various control measures. In that case, following practices may be adopted for an effective control of pests.

1. Start with a thorough understanding of all pesticidal life stages especially the most vulnerable stages
2. Start with a thorough understanding of different types of pests based on their feeding habit
3. Comprise of regular monitoring and sampling of pests to identify the pest
4. Comprise of an integrated pest management plan based on the understanding of the pest nature

Some of the major insects that are found affecting garden plants as well as container-grown plants are, aphids, mealy bugs, thrips, whiteflies, fruit flies and shore flies, gnats, scales, and various types of mites such as spider mites, cyclamen mites, and broad mites. Aphids are soft bodied small insects that normally infest terminal buds and lower leaf surfaces. Aphid infestation can be detected by the presence of sooty mold and stunted plant growth.

IPM for the control of aphids includes use of natural enemies such as ladybeetles and parasitic wasps; use of wire-meshed windows and doors for indoor-grown plants; and sanitation of the growing premises.

Mealy bugs are small insects with their bodies covered with white wax-like material. They are normally present on the lower side of the leaves, base of the leaf stems and along the veins as white cottony mass. Mealy bug infestation can be detected by yellowing and stunting of the plants, and by the presence of sooty mold.

IPM of mealy bugs comprises of various pest control practices such as regular inspection of plants for mealy bug infestation; destruction of infested plants; use of insecticidal sprays; use of insecticidal growth regulators if beneficial insects are present and rotation of insecticides.

Scales are small scale-like crawling insects present on the lower surfaces of leaves and stems. Scale infestation can be detected by the presence of honey dew on the plants and yellowing of the foliage. IPM of scales include regular monitoring; removal of infested plants; and use of insecticidal soaps.

Thrips are small insects that normally infest flower parts and leaves. Infested buds fail to open and flowers turn brown. IPM comprises of field sanitation, use of sterilized growing media, use of screened doors and windows for indoor plants; and rotation of chemical insecticides.

Gnats and shore flies are winged insects that breed in stagnant water; normally present in the wet growing media. IPM comprises of drainage of excess water; removal of infested growing media; and avoiding over watering and over fertilization. Rotations of chemical insecticides are also practiced.

Infestation of spider mites is common during hot, dry conditions. They feed on lower surfaces of leaves and hence they are normally present on lower sides of leaves. IPM comprises of regular monitoring and rotation of chemical insecticides.

Cyclamen mites are microscopic insects normally present inside flower buds. Infestation is high during low temperature, high humidity environment. IPM comprises of hot water bath of infested plants and use of chemical insecticides.

Broad mites normally infest flower parts and foliage and normally present on the lower surfaces of leaves and inside of flowers. IPM comprises of use of chemical insecticides, practicing field sanitation and elimination of infested plants.

Integrated Weed Management: Similarly, weeds of jalapeno plants can effectively be controlled by adopting integrated weed management practices. Major preventive measures that can be adopted for weed control are, summer deep ploughing to expose and destroy the underground vegetative parts of the rooted perennial weeds and by following the recommended agronomic management practices of land preparation, planting distance, fertilizers and irrigation to have healthy plants. Smoothing of weeds by using mulching with straw/hay/plastic sheets etc may be a good control practice. Hoeing and weeding may be done regularly by using hand operated implements for controlling weeds as and when required.

Integrated Disease Management: Integrated disease management (IDM) practices are similar to that of IPM practices. That is, IDM uses a combination technology of cultural, mechanical, biological and chemical control of diseases. When IPM is taken care of well, then disease incidences in the growing fields will be very less as most of the diseases are spread by insects-pests only.

Remember, in organic growing, emphasis is on "prevention" rather than "curing".

Control of Soil-Borne Plant Diseases:

Tillage at the time of field preparation exposes eggs and larvae of soil borne pathogens and exposing the tilled field for some time to the direct sunlight controls growth of these pathogens. *Soil solarisation* and soil drenching with permitted fungicides prevent the infestation of soil-borne pathogens up to a great extent. *Soil fumigation* is another important practice to be adopted for controlling soil-borne diseases. Using *trap crops* such as marigolds along the boundaries of the fields may prevent nematode attack and diseases caused by them.

Control of Seed-Borne Plant Diseases:

Use of certified organic seeds and seedlings as planting materials and treating these planting materials with a recommended organic fungicide controls incidences of seed-borne plant diseases.

In case of disease incidences, permitted fungicides like Bordeaux mixture (1%) and pyrethrum-based fungicides may be used as a control measure.

HARVEST AND POST HARVEST PRACTICES

While harvesting jalapenos, the following points may be taken into consideration.

Harvest the fruits at the correct maturity stage. It takes about three to four months for the transplanted seedlings to grow into full maturity and produce ripe fruits. That is, within a span of 90-120 days, a grower can harvest his/her jalapeno pepper crop.

Harvest fresh jalapenos when they are bright green in color and become firm and crisp. Ideal size of a ripe, green jalapeno pepper is about 4 - 6 inches long.

For processing purposes, colored jalapenos should be harvested. Correct maturity stage for harvesting colored jalapenos is when the pod of a green jalapeno pepper has completely turned into full bright red or orange color.

Harvest the fruits at right time. First harvest is done about 3 months after transplanting. That is, when the fruits reach a size of 3 to 4 inches in length. After that jalapeno plants continue to bear for two to three months more. Harvest up to three or four times during the season. Harvest each fruit carefully along with the stem by using a sharp knife or blade while avoiding fruit injuries.

Yield: Generally a healthy Jalapeno plant yields approximately 1–1.5 kg per plant. There are reports about small scale producers of Jalapeno peppers in Hawaii islands being able to harvest 2,000 to 6,000kg per hectare whereas in USA, under intensive cultivation practices, the yield, in certain cases, had gone up to40,000 kg/ha.

Shelf life: At proper temperature and humidity, shelf life of jalapeno peppers is two to three weeks.

Quality Indices for Raw Green Jalapenos

According to UC Davis, the following quality indices may be considered for assessing the quality of jalapenos.

Size, shape and color of jalapeno peppers: All jalapenos should have uniform shape, size and color typical of that particular variety.

Fruit Texture and Firmness of Jalapenos- All jalapenos should have a firm fruit texture. Fruit Skin should have glossy appearance.

Other quality parameters to be considered- Jalapenos should be free from defects such as cracks, decay, and sunburn. They should be free from mechanical damage also.

Pre-Cooling Requirement: Freshly harvested Jalapeno peppers should be cooled as soon as possible to reduce moisture loss. They are not as chilling sensitive as bell peppers. Two precooling options that can be explored are Room Cooling and Forced Air Cooling.

Postharvest Cleaning of Fruits: Chlorinated water @ 75 to 100 ppm is ideal for cleaning freshly harvested fruits.

Grading Standards: There are no recognized international grading standards for jalapenos.

Packing: Jalapeno peppers are generally packed in carton boxes or baskets. They are generally packed loosely in 10 pound carton boxes. Large carton boxes may also be used if central dividers are used. Weight of a standard carton is 12–15 Kg. However caution is required not to use top icing on packed Jalapeno peppers as the ice may cause water spots on peppers.

Storage: Jalapenos may be stored in clean and transparent

polybags in refrigerators for home uses. For industrial processing, cold storage up to 2 -5weeks is recommended. Jalapeno peppers have the same storage requirements as beans, cucumber, eggplant, bell peppers, and squash. Hence these vegetables may be stored together without detrimental effect.

Optimum Temperature: Jalapeno peppers suffer from moisture loss, shriveling, color changes and decay when they are stored at a temperature above 7.5°C (45°F). At 7.5°C (45°F), maximum shelf-life (3-5 weeks) is obtained. At 5°C (41°F), peppers may be stored for at least 2 weeks without visible signs of chilling injury. But after 2 weeks, chilling injury is detected. Symptoms of chilling injury include pitting, decay, discoloration and softening. Colored peppers are less chilling sensitive than green peppers. Freezing temperature of fresh jalapeños is 31°F (Source: UC Davis).

Optimum Relative Humidity: At a relative humidity (RH) of more than 95%, jalapeno peppers can be kept at their best quality.

Rates of Respiration: Respiration rate of jalapenos at various temperatures is given below:

Temperature	ml CO_2/kg·hr
10°C (50°F)	5-10
20°C (68°F)	20-30
27°C (81°F)	40-80

Source: UC Davis (To calculate heat production multiply ml CO_2/kg·hr by 440 to get BTU/ton/ day or by 122 to get kcal/metric ton/day)

Ethylene Production and Sensitivity : Jalapeno pepper is a non-climacteric fruit and therefore produce very low levels of ethylene, i.e.0.1-0.2 µl/kg·hr at 20-25°C (68-77°F).Jalapeno peppers do not respond to ethylene treatment.

What are climacteric and non-climacteric fruits?

Climacteric fruits produce high levels of ethylene (ripening hormone) during storage while non-climacteric fruits do not show much ethylene producing activity during storage.

Controlled Atmosphere Storage: At 7-8°C of storage temperature, CA (controlled atmosphere) of 3-5% O_2 in combination with 0-5% CO_2 provides only slight benefit to Jalapeno peppers. Low O_2 atmospheres may retard color development while high CO_2 atmospheres (>5%) may cause pitting, discoloration, and softening. (Source: UC Davis).

Pathological Disorders in Jalapeno Peppers

Botrytis or grey mold decay is a fungal infection. Field sanitation and prevention of injuries on the fruits help reduce its incidence. High CO_2 levels (>10%) during storage and hot water dips of peppers (at 55°C for 4 minutes) also control botrytis rot.

Alternaria rot is also a fungal disease that appears on the stem end of the fruit. Field sanitation and hot water treatment of peppers may reduce the infection up to a great extent.

Bacterial soft rot is a bacterial infection that results in the decay of fruit tissues. This is mainly because of the excess moisture presence in the peppers as a result of washing and cooling process. Therefore bacterial soft rot is common on washed or Hydro cooled pepper fruits. Best control measure is to avoid fruit injuries during washing and post-harvest handling. Also care should be taken to dry washed peppers properly before further processing.

Physiological Disorders in Jalapeno Peppers

Blossom end rot occurs as a slight discoloration or a severe dark sunken lesion at the blossom end; it is caused by temporary insufficiencies of water and calcium. It may also occur under high temperature conditions when the peppers are rapidly growing

Chilling injury occurs at very low temperatures. Symptoms of chilling injury include surface pitting, water-soaked areas and discoloration of the seed cavity

In addition to the disorders that are mentioned above, there are other postharvest defects also that are affecting jalapeno peppers. These are mechanical damage due to crushing, stem bruises, cracks, etc and weight loss due to moisture loss and poor quality due to storage decay.

PROCESSING OF JALAPENOS

There is a huge demand for various processed and value-added products of jalapenos in the global processed food markets. Major processed jalapeno products available in the market are Chipotle, canned jalapenos, quick frozen jalapenos (Jalapeno IQF), jalapeno mash, and jalapeno sliced peppers.

Figure 16: Chipotles

Jalapenos are easily perishable and should be processed and preserved in many ways for long-term usage. Chipotles are actually smoked jalapenos. Canning is a popular method of preserving jalapenos. Both sliced and chopped jalapenos can be canned. Sometimes, whole jalapeno peppers are also used for canning purposes. The process of sealing foodstuffs hermetically in containers after sterilizing them by heat for long storage is known as canning. IQF jalapenos are nothing but frozen jalapeno peppers. Jalapenos can be pickled and jellied also for future use.

Processing Jalapenos to Chipotle

Smoked or dried jalapeno peppers are often termed as *chipotle*. Fresh, ripe red jalapenos are selected, cleaned and de-stemmed before subjecting them to coring, drying and smoking process. Chipotle is characterized by a strong, spicy smoke flavour. Scoville heat unit of dried and dark red chipotle is is greater than fresh green jalapenos i.e. between 20,000 - 35,000 SHU. That means, chipotle peppers are hotter than fresh jalapenos. Chipotle is an essential ingredient for almost all Mexican dishes and is an excellent seasoning agent.

SHU for Chipotles: SHU for various categories of chipotles that are available in the market are as follows:

1. Mild chipotle-< 25,000 SHU

2. Medium chipotle-> 25,000 SHU

Sizing: Whole, de-stemmed with 1 ¼" – 3 ½" in Length and ¾" – 1 ¼" in Width

Microbiological Properties: Major microbiological properties of chipotle are given below:

1. Total Plate Count (TPC)- 2,000,000 cfu/ml max

2. Coliform-500 cfu/ml max

3. E. Coli- Negative

4. Salmonella- Negative

5. Yeast / Mold-500 max

Foreign materials- Insects, metal particles, glass; wood, plastic, stones, paper, and hair shall be absent

Chemical and Physical Properties: Major chemical and physical properties of chipotle are given below:

1. Moisture-12% Max

2. Sulfites- None added

Packaging: Chipotle is generally packed in corrugated cardboard boxes of various sizes and in various quantities.

1. Whole Dried Chipotle-30 Lb / Box

2. Chipotle dices-44 lb (20 kg) / Box

3. Chipotle granules-44 lb (20 kg) / Box

4. Chipotle powder 40 mesh-55 lb (25 kg) / Box

Market Price: Price of whole dried chipotle for export markets is at US $3 - 3.25 / Pound (lb) as on January 2013.

Optimum Storage Conditions: Chipotle should be stored in cold, dark and dry area at 70 degrees F or below.

Shelf life: Up to 18 months under optimum storage conditions.

Processing Jalapenos to Canned Jalapenos

For canning purposes, freshly harvested good quality jalapeno peppers are used. During the first step of the canning process, peppers are sorted, washed, cleaned and air dried. During the second step, peppers are trimmed, sliced and washed again. Whole jalapeno peppers may also be canned. Third step is blanching.

Sliced peppers are blanched for 3-4 minutes in water containing 2% salt, 0.1% citric acid and 100 milligrams of ascorbic acid. Fourth step is filling the containers such as cans, bottles and glass jars. Water is drained and pepper slices are filled into the containers containing 2% hot brine solution (composition of this solution is 0.1% citric acid and 100 milligrams of ascorbic acid). Fifth step is sterilization. Containers containing pepper slices are pasteurized at 85 degree Celsius for 5-7 minutes to sterilize them. Final step is sealing of containers and sterilization of sealed containers. Sealed containers are subjected to sterilization process at 116 degree Celsius for 35 minutes. Sterilized containers are stored at room temperature in a cool, dry place.

An overview of canning process of jalapeno peppers is given below:

Step 1:   Selection of Freshly Harvested, Good Quality Jalapenos
Step 2   Sorting and Grading Process
Step 3   Washing and Cleaning; Air Drying
Step 4   Chopping or Slicing
Step 5   Blanching
Step 6   Filling in Containers
Step 7   Brining
Step 8   Pasteurisation and Sterilization of Containers
Step 9   Sealing of Containers
Step 10   Processing and Final Sterilization of Sealed Containers
Cooling and Storing at Room Temperature

Nutrition in Canned Jalapenos (solids and liquids)

According to the USDA nutrient database, nutrition in Canned Jalapenos (solids and liquids together) is as follows:

Nutrient	Unit	Value per 100.0g
Water	g	88.89
Energy	kcal	27
Protein	g	0.92
Total lipid (fat)	g	0.94
Carbohydrate, by difference	g	4.74
Fiber, total dietary	g	2.6
Sugars, total	g	2.14
Calcium, Ca	mg	23
Iron, Fe	mg	1.88
Magnesium, Mg	mg	15
Phosphorus, P	mg	18
Potassium, K	mg	193
Sodium, Na	mg	1671
Zinc, Zn	mg	0.34
Vitamin C, total ascorbic acid	mg	10
Thiamin	mg	0.043
Riboflavin	mg	0.038
Niacin	mg	0.403
Vitamin B-6	mg	0.19
Folate, DFE	µg	14
Vitamin A, IU	IU	1700
Vitamin E (alpha-tocopherol)	mg	0.69
Vitamin K (phylloquinone)	µg	12.9
Fatty acids, total saturated	g	0.097
Fatty acids, total monounsaturated	g	0.053
Fatty acids, total polyunsaturated	g	0.514

Similarly, nutrient composition of **canned, whole jalapenos** (weighing 22g) is given below:

Nutrient	Unit	Pepper 22g
Water	g	19.56
Energy	kcal	6
Protein	g	0.2
Total lipid (fat)	g	0.21
Carbohydrate, by difference	g	1.04
Fiber, total dietary	g	0.6
Sugars, total	g	0.47
Calcium, Ca	mg	5
Iron, Fe	mg	0.41
Magnesium, Mg	mg	3
Phosphorus, P	mg	4
Potassium, K	mg	42
Sodium, Na	mg	368
Zinc, Zn	mg	0.07
Vitamin C, total ascorbic acid	mg	2.2
Thiamin	mg	0.009
Riboflavin	mg	0.008
Niacin	mg	0.089
Vitamin B-6	mg	0.042
Folate, DFE	µg	3
Vitamin A, IU	IU	374
Vitamin E (alpha-tocopherol)	mg	0.15
Vitamin K (phylloquinone)	µg	2.8
Fatty acids, total saturated	g	0.021
Fatty acids, total monounsaturated	g	0.012
Fatty acids, total polyunsaturated	g	0.113

A comparative analysis of nutrients present in canned, chopped jalapenos and canned, sliced jalapenos may be done by using the table below:

Nutrient	Unit	Chopped (136g cup)	Sliced(104g cup)
Water	g	120.89	92.45
Energy	kcal	37	28
Protein	g	1.25	0.96
Total lipid (fat)	g	1.28	0.98
Carbohydrate	g	6.45	4.93
Dietary Fiber	g	3.5	2.7
Sugars	g	2.91	2.23
Calcium, Ca	mg	31	24
Iron, Fe	mg	2.56	1.96
Magnesium, Mg	mg	20	16
Phosphorus, P	mg	24	19
Potassium, K	mg	262	201
Sodium, Na	mg	2273	1738
Zinc, Zn	mg	0.46	0.35
Vitamin C	mg	13.6	10.4
Thiamin	mg	0.058	0.045
Riboflavin	mg	0.052	0.04
Niacin	mg	0.548	0.419
Vitamin B-6	mg	0.258	0.198
Folate, DFE	µg	19	15
Vitamin A	IU	2312	1768
Vitamin E	mg	0.94	0.72
Vitamin K	µg	17.5	13.4
Saturated Fatty acids	g	0.132	0.101
*MUFA	g	0.072	0.055
**PUFA	g	0.699	0.535

*MUFA-Mono Unsaturated Fatty Acids /**PUFA- Poly Unsaturated Fatty Acids

Processing Jalapenos to IQF Jalapenos

Jalapeno IQF is prepared by using Individual Quick Freezing (IQF) Technology where individual whole jalapeno fruit is cleaned, washed, and dried before it is quick frozen to prolong its shelf life. IQF jalapenos are packed in poly laminated bags and then bulk packaged in cardboard boxes of convenient sizes in desired quantities (i.e. 5 kg, 10 kg etc).

Optimum Transport and Storage Conditions: A list of basic conditions for optimum transport and storage of IQF jalapenos is given below:

1. Transport Temperature-18°C
2. Storage Temperature-18°C
3. Shelf life-12 months

Quality Indices for IQF Jalapenos: A list of quality indices to be considered for the quality assessment of IQF jalapenos is given below:

1. Color- Green / Red
2. Appearance- Whole pepper
3. Aroma, flavor- Hot flavor typical of freshly harvested jalapeño
4. Pungency- Hot

Physical Properties: A list of physical properties to be considered for top quality IQF jalapenos is given below:

1. Size-1.25 – 3 inches

2. pH-5.8 - 6.3

3. Defects-< 0.5% by weight of off cut pieces; < 0.5% by weight off color pieces

4. Other Desired Properties- No whole insects; No foreign materials

Microbiological Properties: A list of microbiological properties to be considered for top quality IQF jalapenos is given below:

1. Plate count-<10,000 cfu/g

2. Molds and yeasts-<5,000 cfu/g

3. Clostridium botulinum-<1,000 cfu/g

4. Foreign materials in 100 g-Nil

Processing Jalapenos to Jalapeno Mash

Jalapeno mash is prepared by grinding fresh fruits of green and/or red Jalapenos to a paste with salt and then fermented in brine. Before using the jalapenos for mash preparation, jalapeno fruits should be cleaned and should be free from inert matter, extraneous substances etc. Jalapeno mash is normally packed in double layered plastic bags inside a wooden or metal container. Recommended packaging sizes for long distance markets are 200 kilograms, 500 kilograms or 1000 kilograms.

Optimum Transport and Storage Conditions: A list of basic conditions for optimum transport and storage of jalapeno mash is given below:

1. Transport and storage temperature- Room temperature; during transport, protect the products from humidity, dust and direct sun light

2. Shelf life-24 months

Precaution- Avoid any possibility of contamination or damage to the product

Quality Indices for Jalapeno Mash: A list of quality indices to be considered for the quality assessment of jalapeno mash is given below:

1. Color- Dark green/ Red. Color not standardized

2. Appearance- Paste with seeds

3. Aroma, flavor- Hot flavor

4. Pungency- Hot

Physical and Chemical Properties: A list of physical and chemical properties to be considered for top quality jalapeno mash is given below:

1. Soluble solids Brix-20-26.5

2. Total solids %-23-27

3. Pepper solids %-10.5-13%

4. Acetic acidity %-0.5-1.5

5. Salt %-10-14

6. Consistency Bostwick, cm/15 seg-<8

7. pH-<4

Microbiological Properties: A list of microbiological properties to be considered for top quality jalapeno mash is given below:

1. Plate count-<10,000 cfu/g

2. Molds and yeasts-<5,000 cfu/g

3. Clostridium botulinum-<1,000 cfu/g

4. Foreign materials in 100 g- Nil

Culinary Uses: Jalapeno mash is used as a base for salad dressings and marinades.

Processing Jalapenos to Nacho Sliced Peppers

Jalapenos Nacho sliced peppers are prepared by slicing good quality, clean jalapenos in equal sizes (seeds are sifted out). Then jalapeno slices are immersed in hot vinegar at 75°C (167°F) for 5-10 minutes before cooling it. Then preservatives such as sodium benzoate and calcium chloride are added and the finished product is packed. Jalapenos Nacho sliced peppers are packed either in pails or in pouches. Jalapenos nacho pails are packed in HDPE (high density polyethylene) bags of convenient sizes and Jalapenos nacho pouches are packed in HDPE transparent pouches of various sizes.

Optimum Transport and Storage Conditions: A list of basic conditions for optimum transport and storage of jalapeno nacho sliced peppers is given below:

1. Transport and storage temperature- Room temperature; during transport, protect the products from humidity, dust and direct sun light

2. Shelf life-24 months

Precaution- Avoid any possibility of contamination or damage to the product

Recommended Quality Indices: Important quality indices to be considered for jalapenos nacho sliced peppers is given below:

1. *Color- Green*

2. *Appearance- Slices (6 mm thickness)*

3. *Aroma, flavor- Hot flavor*

Physical and Chemical Properties: Major physical and chemical properties to be considered for jalapenos nacho sliced peppers is given below:

1. *Acidity %-1.3-1.7*

2. *Salt %-1-2*

3. *pH %-2.5 -3.5*

Culinary Uses: Jalapeno Nachos are used in continental food cuisines, Mexican food preparations, pizzas, and meat preparations.

Nutrition in Jalapenos Nacho Sliced Peppers

According to the USDA nutrient database, nutrition in Jalapenos Nacho Sliced Peppers is as follows:

Nutrient	Unit	Value/100.0g
Water	g	91.3
Energy	kcal	13
Carbohydrate	g	3.33
Fiber, total dietary	g	3.3
Sugars, total	g	3.33
Calcium, Ca	mg	67
Sodium, Na	mg	1000
Vitamin C	mg	12
Vitamin A, IU	IU	333

In other words, a serving of 30g of jalapenos nacho sliced peppers contains the following nutrient composition:

Nutrient	Unit	Serving of 30g
Water	g	27.39
Energy	kcal	4
Carbohydrate	g	1
Fiber, total dietary	g	1
Sugars, total	g	1
Calcium, Ca	mg	20
Sodium, Na	mg	300
Vitamin C	mg	3.6
Vitamin A, IU	IU	100

Processing Green Jalapeno Slices in Brine

Good quality green jalapenos are selected, cleaned and sliced. Slices are then subjected to a controlled maturation process before filling them in polyethylene or polyester pouches containing brine solution. These pouches are bulk packaged in metal or wooden containers of large sizes. Bulk packages are available in 200 kilograms, 500 kilograms and 100 kilograms.

Optimum Transport and Storage Conditions: A list of parameters to be considered for optimum storage and transport of jalapenos preserved in brine is given below:

1. Transport and storage temperature- Room temperature; during transport, protect the products from humidity, dust and direct sun light

2. Shelf life-24 months

Precaution- Avoid any possibility of contamination or damage to the product

Recommended Quality Indices: Major quality indices to be considered for jalapenos in brine are given below:

1. *Color- Green*

2. *Appearance- Slices (6 mm thickness)*

3. *Aroma, flavor- Hot flavor*

Physical and Chemical Properties: Major physical and chemical properties to be considered for jalapenos in brine are given below:

1. *Acidity %-0.5-1.0*

2. *Salt %-7-8*

3. *pH %-3.5 – 4*

CONTAINER GROWING OF JALAPENOS

The practice of growing plants in pots and containers is known as container gardening. Container gardening is mostly practiced in urban homes, multistorey buildings of towns and cities and in places where availability of land is a major constraint. Jalapeno peppers are a favorite container-grown plant among urban population and this plant can be grown in containers successfully.

Some of the major considerations while preparing for raising a container garden of jalapeno plants are:

1. Choosing right containers
2. Choosing a good growing media
3. Preparing a gardening schedule or a garden calendar
4. Arranging all necessary gardening tools

Selection of Suitable Containers/Pots: Earthenware pots are most commonly used containers for growing jalapeno plants. Wooden barrels and planters can also be used and these containers should be painted from inside as well as outside with waterproof oil-paints before using them. Plastic jars, pots, dishes and bowls, glazed clay and china (porcelain) pots, shallow bowls and troughs, pottery containers, boxes and crates, cement pots, cans and buckets, tin boxes, drums, brass and copper containers may also be used according to the circumstances and growing requirements.

While it comes to the shape of the containers, any shape of the container can be used whether it is circular, round, oval, elliptical, cone, pyramid, rectangular, square, or heart-shaped.

Containers should have at least one hole of an adequate size at the bottom as in earthen pots, to drain out excess water. Containers can easily be placed on the terrace, window sills, window boxes, balcony and verandah where sunlight is available for the plants. Containers can hold sufficient volume of growing media and should be lightweight and easy to handle. Containers should be durable and free of toxic substances and also should prevent root circling.

Selection of Suitable Growing Media or Soil and Fertilizers: Growing media may be prepared from a mixture of good soil, river sand, well-decomposed organic manure (compost or farmyard manure), nitrogenous fertilizers (urea or ammonium sulphate) and recommended insecticides and fungicides. Growing media should be able to hold seedlings firmly. Media should be free of insect-pests and diseases and should have good water-holding capacity. Growing media should also have excellent aeration and drainage.

How to prepare a suitable growing media?

Mix good soil, river-sand and well-rotten organic manure in equal quantities with the help of a khurpi or shovel. Make sure that the mixture is free from various soil-borne insects, termites, red ants and cut worms. Add a small quantity of recommended fungicide (organic pyrethrum-based fungicides may be used)to the mixture before filling it in the containers; this helps to prevent seedling rotting caused due to fungal infections. After raising a crop for one season, the container mixture should be removed and cleaned of roots and exposed to the sun for a few days. This growing media could then be reused after mixing one-third the quantity of organic manure and a small quantity of recommended fungicide.

Alternatively, growers can prepare their own compost or vermicompost by using kitchen wastes and use it as growing media for their container-grown plants.

Preparation of a Garden Calendar: If you are planning for container-growing of jalapeno plants, then you should prepare a gardening schedule beforehand. A garden calendar should clearly indicate the following parameters,

1. Sowing Time: When to sow the seeds?
2. Fertilizer Schedule: When to fertilize plants? How many times?
3. Irrigation/Watering Schedule: How much irrigation is required? And when?
4. Weeding and Aftercare: What are the weed control measures to be adopted?
5. Harvesting Time: When can fruits be harvested?

Arranging Garden Tools: A container garden needs to have some garden tools such as shovels, garden knife, measuring sticks, strings, hand hoes, dusters, sprayers, watering cans, and hand cultivators.

Figure 17: Container Gardening

How to Grow Jalapeno Plants in Containers?

After arranging containers, growing media and necessary garden tools, a grower can start raising seedlings of jalapeno plants. Jalapeno plants are propagated by seeds.

Favourable temperature for seeds germination is 15–30°C. Seedlings are raised in well-prepared nursery beds. Nursery beds are prepared in a well-shaded area and seeds are either broadcasted or sown in rows in nursery beds. Germination process is slow and it may take up to 12 weeks for the seeds to germinate. Regular light watering is necessary to keep the nursery bed moist always. Moist soil facilitates quick germination of the seeds and seedlings will become ready for transplanting when they reach up to 6-8 weeks old.

For growing in containers, transplanting of seedlings can be done at any time. While transplanting, a single healthy seedling may be transplanted in each container. Sometimes two or more seedlings may be transplanted if necessary. Remember that plants in pots and containers need a lot of care and attention. So it is essential to water frequently depending on the season, kind of crop and size of the plant and container. Plants need extra water in dry summer season, so watering should be done twice a day (morning and evening). Too much watering can be as harmful in winter as too little in summer. In the rainy season, proper water drainage is essential if plants are placed in the open areas. If there is heavy rain, containers should be tilted slightly to drain out the excess water from the top.

Topdressing with nitrogenous fertilizers improves plant growth and yield. This can be done by foliar application of liquid nitrogen, urea or ammonium sulphate in small quantities. Alternatively, urea granules@ 5–10g/container may be applied in moist soil once a week or 10 days, starting from 2 weeks after transplanting seedlings. High dose of fertilizer is very harmful since it can kill the plants.

If urea or ammonium sulphate is applied in dry soil, the plants must be watered immediately. Young plants may require staking. Hand-hoeing and weeding with the help of a small khurpi should be done periodically to remove weeds. Weeds should also be uprooted gently by hand.

For container-grown plants, water requirement may be determined by weighing pots, feeling growing medium and by using indicator plants that readily show water stress. Watering should be done in early morning to minimize evaporation loss. Applying water in two or more applications conserves water.

Major insects found in container-grown plants are, aphids and jassids, and fruit flies and fruit borers. Aphids and jassids are small-sucking insects, injuring the plants especially in early stage of their growth. Fruit flies and fruit borers damage young fruits and make them unfit for consumption. Use of organic insecticides such pyrethrum-based sprays or tobacco emulsions, or neem oil based solutions effectively controls these insects. Use of mechanical traps (colour traps, light traps etc) and manual picking of insects may also be tried for insect control. After spraying with insecticides, vegetables should not be harvested for 7 days for consumption. Fungal diseases (damping off and wilt) and viral diseases affect the plants particularly in the rainy season. Fungal diseases can be controlled by drenching the soil with an appropriate fungicide. Virus affected plants should be removed and destroyed.

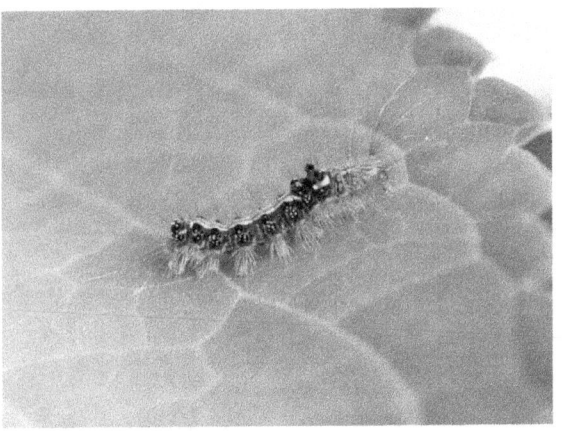

Figure 18: Leaf-Eating Caterpillars

Flowering begins approximately after 8 weeks of transplanting. It takes about three to four months for the transplanted seedlings to grow into full maturity and produce ripe fruits. That is, within a span of 90-120 days, a grower can harvest his/her jalapeno pepper crop.

Pruning is essential to remove root circling when root systems become too long for their containers. Root circling can be prevented by air root pruning; by using bottomless containers and by using copper compounds. Pruning may be necessary to induce new growth and to remove undesirable growth and damaged or dead stems.

MARKETING OF JALAPENOS

Jalapeno peppers may be marketed both in local and export markets. There is a strong local demand for jalapenos and chipotles in Mexico and Latin American countries. Other major markets for jalapenos are in USA, EU, Middle East nations and South Asian countries.

Nowadays, large-scale producers are exploring branding options for their produce so that they can brand their product and build a reputation in the markets based on the quality of their products. For branding and export marketing, a grower may need US FDA (U.S. Food and Drug Administration) or any other relevant food quality certifications such as Kosher food certification.

Local Markets

Market freshly harvested peppers as soon as possible in order to avoid quality deterioration and moisture loss. Study local market before marketing your product for local or national markets. It is always advisable to know local preferences for certain varieties of jalapeno peppers before growing them for local markets.

Cold Storage

Transport freshly harvested jalapenos from fields to pack houses as soon as possible so that precooling can be done at the earliest to remove field heat from the produce; precooling is followed by other post-harvest processes such as packing, labeling and storage at optimum conditions

Export Markets

Mode of transport for exporting jalapenos is by air or by sea. For export by air, product is transported the day after harvest. Export by sea is feasible when travel time is less than 14 days and product is kept at temperatures of 12–13°C. Each product package should have appropriate labels and each label should have information pertaining to product, packing, weight, grade, producer, etc. Major export markets for Jalapeno peppers are Canada, Europe, Russian Federation and the USA. Presence of corky striations on the fruit surface is considered unattractive in the U.S. market. Mexican market prefers red peppers or mature green peppers. Jalapeno peppers have a well-established global market for growers from Asia, Mexico, and Central America.

Economics of Production

Cost of production of peppers varies depending upon locality, soil type, variety and other relevant parameters. Major production expenses include cost of seeds; labor cost involved in field preparation and planting process, and costs of cultural management; cost of fertilizers and pesticides, manpower costs involved in field management activities such as irrigation management, disease-pest management and weed management; costs of harvesting process and postharvest management activities. Major income obtained is through the sales of fresh peppers and processed peppers. In global markets, a box of 4.5 Kg of fresh Jalapeno peppers is sold at a price of $10 - $12 (as on Jan 2013) during seasons.

Risks Involved

Three major risks faced in all cases of agricultural production are Production-related risks; Price-related risks and Finance-related risks. Risks associated with the production are unexpected natural calamities, insect-pest damage and disease incidences. Price risks include unexpected fall in product prices while major finance-related risks are sudden increase in loan interest rates.

Growers are advised to be well-prepared to manage these risks before embarking on a large-scale commercial production of jalapenos. They can minimize these risks by gathering as much information as possible on production practices, product price trend over a period of time and other relevant market information.

Bibliography

Germplasm Resources Information Network . (2013). Retrieved , 2013 , from GRIN NPGS: http://www.ars-grin.gov/npgs/

UC Davis Post Harvest. (2013). Retrieved, 2013, from UC Davis Post Harvest Technology Center: http://postharvest.ucdavis.edu/producefacts/

USDA ARS . (2013). Retrieved, 2013 , from USDA Agricultural Research Service : http://www.ars.usda.gov/main/main.htm

USDA Nutrient Database . (2013, December Thursday). Retrieved, 2013, from USDA: http://ndb.nal.usda.gov/ndb/search/list

USDA Plant Database . (2013). Retrieved , 2013, from USDA Plant Database : http://plants.usda.gov/java/

Roby Jose Ciju

ABOUT THE AUTHOR

Roby Jose Ciju is the author of *'The Art of Perfect Living: The 7 Personal Powers for Perfection'*, an inspirational book based on scriptural wisdom. She is also a professional horticulturist and an agribusiness consultant with a Masters Degree in Horticulture and a Post Graduate Diploma in Agri-Supply Chain Management.

She has founded www.agrihortico.com, a website dedicated for publishing information on Food & Agriculture Topics. She has written more than 40 books on various Food & Agriculture topics till date and her best-selling books are, Mushroom Farming, Moringa, Curryleaf, Jalapeno Peppers, and Growing Ginger, Turmeric and Arrowroot. She may be contacted at roby@agrihortico.com. You may follow agrihortico at https://twitter.com/agrihortico1. Her official website is available at www.robyjoseciju.com.

Roby Jose Ciju